NIST Special Publication 260-138

Standard Reference Materials:

Tin Freezing-Point Standard — SRM 741a

Gregory F. Strouse

Chemical Science and Technology Laboratory
Process Measurements Division
National Institute of Standards and Technology
Gaithersburg, MD 20899-0001

and

Natalia P. Moiseeva

D.I. Mendeleyev Research Institute of Metrology
Thermodynamics Laboratory
St. Petersburg, Russia

U.S. DEPARTMENT OF COMMERCE, William M. Daley, Secretary
TECHNOLOGY ADMINISTRATION, Gary R. Bachula, Acting Under Secretary for Technology
NATIONAL INSTITUTE OF STANDARDS AND TECHNOLOGY, Raymond G. Kammer, Director

Issued June 1999

FOREWORD

Standard Reference Materials (SRMs®) as defined by the National Institute of Standards and Technology (NIST) are well-characterized materials, produced in quantity and certified for one or more physical or chemical properties. They are used to assure the accuracy and compatibility of measurements throughout the Nation. SRMs are widely used as primary standards in many diverse fields in science, industry, and technology, both within the United States and throughout the world. They are also used extensively in the fields of environmental and clinical analysis. In many applications, traceability of quality control and measurement processes to the national measurement system is carried out through the mechanism and use of SRMs. For many of the Nation's scientists and technologists, it is therefore of more than passing interest to know the details of the procedures, modes, and philosophy used at NIST to use, produce, and certify SRMs and RMs. The NIST Special Publication 260 Series is a series of papers reserved for this purpose and can be accessed via internet: http://ts.nist.gov/srm.

This 260 publication is dedicated to the dissemination of information on different phases of the preparation, measurement, certification, and use of NIST SRMs. In general, much more detail will be found in these papers than in generally allowed, or desirable, in scientific journal articles. This enables the user to assess the validity and accuracy of the measurement processes employed to judge the statistical analysis, and to learn details of techniques and methods utilized for work entailing greatest care and accuracy. These papers also should provide sufficient additional information so SRMs can be utilized in new applications in diverse fields not foreseen at the time the SRM was originally issued.

Inquiries concerning the technical content of this paper should be directed to the author(s). Other questions concerned with the availability, delivery, price, and so forth, will receive prompt attention from:

>Standard Reference Materials Program
>Bldg 202, Room 204
>National Institute of Standards and Technology
>Gaithersburg, MD 20899-2320
>Telephone: (301) 975-6776
>FAX: (301) 948-3730
>e-mail: srminfo@nist.gov, or
>www:http://ts.nist.gov/srm

>Thomas E. Gills, Chief
>Standard Reference Materials Program

TABLE OF CONTENTS

	PAGE
1. Introduction	2
2. SRM Samples and Filling Procedure of the Fixed-Point Cells	2
2.1 SRM Sample	2
2.2 Tin Fixed-Point Cell	3
2.3 Assembling of a Fixed-Point Cell	3
3. Certification Procedure for SRM 741a	6
3.1 Freezing-point realization	6
3.2 Melting-point realization	7
3.3 Direct comparison measurements	7
4. Analysis of Results	8
4.1 Analysis of SRM tin freezing-point curves	8
4.2 Analysis of SRM tin melting-point curves	9
4.3 Analysis of direct comparisons	9
5. Uncertainties	17
5.1 Uncertainty of direct comparison measurements	17
5.2 Uncertainty assigned to SRM 741a	18
6. Application	18
7. Conclusions	18
8. References	20
9. Appendix A	22
10. Appendix B	24

LIST OF FIGURES

FIGURE NO.		PAGE
1.	Schematic of an tin fixed-point cell assembly	4
2.	Freezing curves for tin fixed-point cell Sn 97-1	10
3.	Freezing curves for tin fixed-point cell Sn 97-2	11
4.	Freezing curves for tin fixed-point cell Sn 97-3	12
5.	Melting curves for tin fixed-point cell Sn 97-1	13
6.	Melting curves for tin fixed-point cell Sn 97-2	14
7.	Melting curves for tin fixed-point cell Sn 97-3	15
8.	Direct freezing plateau comparison results of Sn 97-1, Sn 97-2 and Sn 97-3 with Sn 88A (laboratory standard).	16

APPENDICES

		PAGE
A.	Johnson Matthey emission spectrographic assay of tin metal for SRM 741a.	22
B.	Certificate for SRM 741a: Tin Freezing-Point Standard.	24

Standard Reference Materials: Tin Freezing-Point Standard – SRM 741a

Gregory F. Strouse
National Institute of Standards and Technology (NIST)
Process Measurements Division
Gaithersburg, MD 20899

Natalia P. Moiseeva
D. I. Mendeleyev Research Institute of Metrology (VNIIM)
Thermodynamics Laboratory
St. Petersburg, Russia

Abstract

The freezing point of tin (231.928 °C) is a defining fixed point of the International Temperature Scale of 1990 (ITS-90). Realization of this freezing point is performed using a fixed-point cell containing high-purity (≥99.9999 % pure) tin. A 24 kg single lot of tin (99.999 97 % pure) constituting Standard Reference Material® (SRM®) 741a has been evaluated, and certified as suitable for use in the realization of the freezing-point temperature of tin for the ITS-90. Based on results obtained with three fixed-point cells containing random samples of SRM 741a, the expanded uncertainty ($k=2$) assigned to the freezing-point temperature of the metal is 0.53 m°C. SRM 741a has a higher purity and a smaller expanded uncertainty ($k=2$) than that of SRM 741 of 99.9999 % purity and 1 m°C, respectively. In this document, the methods used for the fabrication of the tin freezing-point cells and for the evaluation of SRM 741a are described.

Disclaimer

Certain commercial equipment, instruments or materials are identified in this paper in order to adequately specify the experimental procedure. Such identification does not imply recommendation or endorsement by NIST, nor does it imply that the materials or equipment identified are necessarily the best available for the purpose.

Acknowledgment

The authors wish to thank the Standard Reference Materials Program for their support in the development of this SRM.

1. Introduction

One of the freezing points of the International Temperature Scale of 1990 (ITS-90) is that of tin (231.928 °C) [1]. This freezing point is realized by using a thermometric fixed-point cell containing high-purity (≥99.9999 % pure) tin metal. Such a fixed-point cell is used for the ITS-90 calibration of standard platinum resistance thermometers (SPRTs) from 0 °C to 232 °C, 0 °C to 420 °C, 0 °C to 661 °C and from 0 °C to 962 °C. To provide evaluated and certified material for this purpose and to replace the depleted stock of SRM 741 (99.9999 % pure), we have developed a Standard Reference Material® (SRM®), the Tin Freezing-Point Standard (SRM 741a).

The certification of SRM 741a was performed by evaluating three fixed-point cells (designated Sn 97-1, Sn 97-2 and Sn 97-3) containing random samples of the single lot of high-purity (99.999 97 % pure) metal constituting the SRM. The certification included the evaluation of freezing and melting curves and the direct comparison with the laboratory tin freezing-point reference cell (Sn 88A) in the NIST Platinum Resistance Thermometry (PRT) Laboratory. Using the three fixed-point cells, these methods of evaluation were used to confirm the purity and the freezing-point temperature of SRM 741a relative to the tin (99.999 99_8 % pure) used in Sn 88A. The fabrication of the tin fixed-point cells used to certify SRM 741a are described and the results from the certification are given.

For the metal to be certified as a freezing-point standard, the fixed-point cells containing samples of the metal should have a freezing-point temperature that is in agreement with the laboratory reference cell containing high-purity metal to within the uncertainties of the measurements and those due to the differences in purity. If the purity of the metal in the new fixed-point cell is greater than that of the reference cell, then it may be of even higher quality, as indicated by being "hotter" than the laboratory standard. A cell that is "hotter" usually has fewer impurities, since impurities contained in the metal fixed-point will usually decrease the freezing-point temperature.

2. SRM Samples and Filling Procedure of the Fixed-Point Cells

2.1 SRM Sample

The high-purity (99.999 97 % pure) tin metal used for SRM 741a was purchased by the NIST Standard Reference Materials Program from Johnson Matthey Company of Spokane, Washington. A 24-kg lot (M6755) of high-purity tin metal, in millimeter size teardrop shot form (nominally 0.014 cm o.d. by 0.018 cm long), constitutes SRM 741a. Johnson Matthey Company placed the randomized metal in plastic bottles sealed in an argon atmosphere, each containing 200 g of tin. The purification of SRM 741A was performed by the Johnson Matthey Company. The emission spectrographic assay of the tin (Appendix A) provided by Johnson Matthey Company shows the total impurity level to be 0.3 µg/g (0.3 parts per million), resulting from 0.1 µg/g of Ag, 0.1 µg/g of Ca and 0.1 µg/g of Si.

2.2 Tin Fixed-Point Cell

Three thermometric fixed-point cells (Sn 97-1, Sn 97-2 and Sn 97-3) were constructed to certify the tin metal for use as an ITS-90 freezing-point standard. Each cell contained 1071 g of the high-purity metal from randomly selected bottles of lot M6755.

As shown in figure 1, the metal (K) was contained within a high-purity graphite crucible (L), with a high-purity graphite cap (I) and a high-purity graphite re-entrant well (J). The graphite assembly was placed inside a precision-bore borosilicate-glass envelope (H) (46.2 cm long, ground to a 5 cm o.d. with a wall thickness of 0.4 cm). Axially located in the annular space between the graphite crucible and the borosilicate-glass envelope was a piece of ceramic fiber blanket (M) (24.3 cm long, 14.1 cm wide and 0.15 cm thick) for thermal insulation between the borosilicate-glass envelope and the graphite crucible and to provide cushioning for the graphite assembly. Above the graphite cap, there was a matte-finished borosilicate-glass guide tube (F) (1.0 cm o.d. with a wall thickness of 0.1 cm), 1.2 cm thick washed-ceramic fiber disks (E) and two graphite heat shunts (G). The heat shunts were placed approximately 3.2 cm and 10.8 cm, respectively, above the top of the graphite cap and were snug fitting in the borosilicate-glass envelope. A space of about 1.8 cm between the top of the borosilicate-glass envelope and the ceramic fiber insulation allows for a silicone-rubber stopper to be glued into place using silicone-rubber sealant in the top of the borosilicate-glass envelope. This silicone-rubber stopper (D) has (1) a modified compression fitting with a silicone-rubber O-ring (C) for inserting and sealing an SPRT into the fixed-point cell, and (2) a stainless-steel gas filling tube (B) (4.3 cm long and 0.3 cm o.d.) for evacuating and backfilling the cell with an inert gas (He). Additionally, the gas filling tube allows for a slight overpressure (0.25 kPa above atmospheric pressure) of the inert gas to prevent contamination of the metal.

The immersion depth of the cells was 18 cm as measured from the sensor mid-point of 25.5 Ω SPRTs to the top of the liquid metal surface (distance from the bottom of the graphite re-entrant well to the top of the liquid level is 20.5 cm). The pressure in the cells during use was 101 325 Pa ±27 Pa.

The high-purity ($\geq$99.9999 % pure) graphite pieces (crucible, well, cap and heat shunts) were purchased from Carbone of America, Ultra Carbon Division, Bay City, Michigan. The usable volume space, allowing for a 1 cm head space between the liquid metal and the underside of the graphite top, is 149 cm^3.

The fixed-point cell assembly of the Sn freezing-point cells was placed inside a three-zone furnace for evaluation. A description of the furnace may be found in Ref. 2.

2.3 Assembling of a Fixed-Point Cell

Any handling procedure of high-purity material is apt to introduce contamination unless extreme care is exercised. The small teardrop shot form of the SRM tin is convenient for handling and filling a crucible during freezing-point cell construction, while a solid cylinder sample might require cutting and cleaning. By using one-time-use polyethylene gloves, every possible effort was made to maintain the purity of the SRM tin and the other fixed-point cell components that come in contact with the tin.

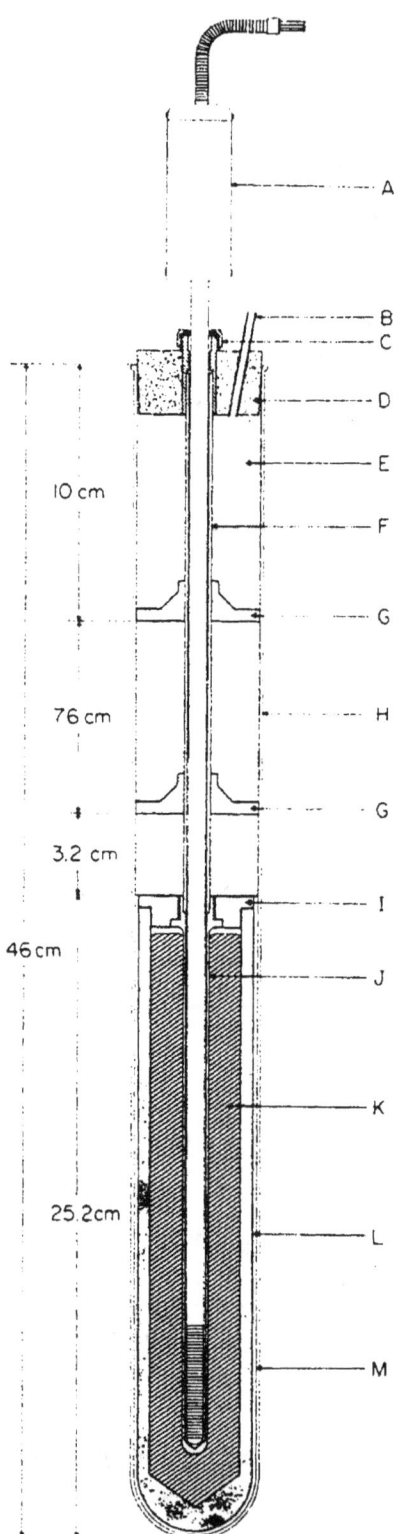

Figure 1. A schematic of a tin freezing-point cell [2] showing: (A) a 25.5 Ω SPRT; (B) fill tube to inert gas (Ar or He) supply and pressure gauge; (C) thermometer gas seal (a modified Swagelok fitting with a silicone-rubber O-ring); (D) silicone rubber stopper; (E) thermal insulation (1.2 cm thick washed Fiberfrax disks); (F) matte-finished borosilicate-glass guide tube; (G) two graphite heat shunts; (H) precision-bore borosilicate-glass envelope; (I) graphite cap; (J) graphite re-entrant well; (K) metal sample; (L) graphite crucible; (M) thermal insulation between the borosilicate-glass envelope and the graphite crucible.

Prior to filling the graphite crucible with the tin metal, the graphite crucible assembly (crucible, cap and re-entrant well) was placed in a silica-glass furnace tube and baked at approximately 600 °C under vacuum overnight. This "bake-out" of the graphite was a final purification to remove hydrocarbons and other contaminants that might have been present from the fabrication process. The vacuum system used during the fabrication of the tin freezing-point cells is an oil-free system consisting of a diaphragm pump and a turbo-molecular pump attached to a gas handling and purification system described in Ref. 2.

The silica-glass furnace tube used in the filling and assembly of the fixed-point crucible is a 4.8 cm o.d. test tube with a silica-glass side-arm pumping tube near the top of the test tube that allows for its evacuation and filling with purified argon. The open end of the furnace tube is sealed with either a solid silicone-rubber stopper or a stopper that has a vacuum seal that allows a silica-glass push rod to be used for pushing the graphite well into place.

After the graphite assembly contained in the glass furnace tube was baked-out and cooled to ambient temperature under vacuum, the furnace tube was purged with purified argon and then the graphite crucible assembly was removed and placed inside a clean polyethylene storage bag. After each use (bake out or fill), the silica-glass furnace tube was (1) cleaned in hot, soapy water; (2) rinsed with copious amounts of water; (3) the inside soaked in 20 % nitric acid-80 % distilled water (volume) for 1 h; (4) rinsed with copious amounts of distilled water, and then (5) dried.

High-purity argon gas is necessary during the construction of the tin fixed-point cell. The gas purification system for argon described in Ref. [2] was designed to remove any hydrocarbons, oxygen and/or water that would contaminate the tin.

In order to introduce 1071 g of tin into the graphite crucible and insert the graphite re-entrant well, two fillings were required. The tin teardrop shots were poured directly into the graphite crucible in air. Approximately 760 g of the shot could be put into the graphite crucible for the first fill. The crucible with the graphite cap and this first filling were placed into the cleaned furnace tube and the furnace tube was placed in the furnace. The system was evacuated for 1 hour, then back-filled with purified argon to a pressure of about 34 kPa, and then evacuated again. This process of pumping, flushing and pumping the system was carried out three times. Upon the final evacuation, the furnace was turned on and the temperature was brought to 235 °C to melt the tin sample. After about 4 hours the sample was completely melted, and the furnace was allowed to cool to ambient under vacuum. Then purified argon was introduced, the graphite crucible was removed from the furnace tube, and the remaining metal was added (about 311 g) to the crucible. The graphite re-entrant well was inserted in the graphite crucible as far as possible through the hole in the graphite cap and the assembly was placed in the furnace tube. A silica-glass push rod extended from the bottom of the graphite re-entrant well through a vacuum seal and for a sufficient distance above the seal at the top of the furnace tube to allow the graphite well to be pushed into place when the tin was molten. Finally, using the method described above, the furnace tube was placed in the furnace, evacuated, flushed with purified Ar, and then evacuated again. Then the metal sample was melted under vacuum. When the sample had melted, the graphite re-entrant well was slowly inserted into the molten metal by pushing it with the silica-glass push rod. When the graphite well was fully inserted, the furnace was turned off, the system allowed to cool to ambient temperature, and the filled graphite crucible assembly removed and placed inside a clean polyethylene storage bag. Later, the filled graphite crucible assembly was placed into its borosilicate-glass envelope and the fixed-point cell assembly that was described earlier. Prior to its use, the borosilicate-glass envelope had been cleaned in the same manner as described for the silica-glass furnace tube.

3. Certification Procedure for SRM 741a

The first step in certifying the tin of SRM 741a was to obtain three freezing and three melting curves for each of the specimens in the cells, using a 25.5 Ω standard platinum resistance thermometer (SPRT). The second step in certifying the SRM metal was to obtain three direct comparisons of the SRM fixed-point cells with the laboratory standard cell (Sn 88A). An Automatic System Laboratories (ASL) F18 30 Hz ac resistance ratio bridge, with a thermostatically controlled (25 °C ±0.01 °C) Tinsley 5685A ac/dc 100 Ω reference resistor, was used to measure the SPRT. A description of the measurement system used in the PRT Laboratory may be found in Ref 3.

3.1 Freezing-point realization

In the realization of the freezing point, the recommended "induced inner freeze" method [4,5] was used. Due to the deep supercool of as much as 25 K below the freezing-point temperature of high-purity Sn, the technique for the realization of the Sn freezing point is different from that of the other ITS-90 freezing-point metals. The freezing point was achieved by heating the cell overnight to approximately 5 K above the freezing-point temperature and then setting the furnace temperature about 0.5 K below the freezing-point temperature of the metal and monitoring the temperature of the metal with the check thermometer (SPRT). When the Sn had cooled to its freezing-point temperature, the freezing-point cell with the check thermometer was removed from the furnace until the start of recalescence. At the beginning of that recalescence, the fixed-point cell was placed back into the furnace, the check thermometer removed and two fused-silica glass rods were inserted for three minutes each into the reentrant well of the cell to induce an inner solid-liquid interface. Finally, the "cold" thermometer was reinserted into the cell and, after equilibrium was obtained, measurements begun. Using an excitation current of 1 mA, thermometer readings were recorded continuously until the freezing was complete. After each freezing-point realization, the SPRT was measured at the triple point of water (TPW) to calculate W(Sn) [W(Sn) = R(Sn)/R(0.01 °C)].

From the analyses of the freezing curve, an estimation of the total impurity level in the metal sample may be made using Raoult's Law of dilute solutions [6-8]. The equation for approximating the effect of impurities on the freezing-point temperature of 100 % pure sample may be written as:

$$\Delta T = \frac{c}{FA} \quad (1)$$

where ΔT is the temperature depression at the fraction F of the sample melted, c is the overall mole fraction impurity concentration and A is the first cryoscopic constant.

Using the total impurity level obtained from the assay of the metal sample and equation 1, an estimation of the depression in temperature at 50 % of the sample frozen may be calculated. This calculated value may then be compared with the experimentally determined value by graphing the freezing curve as a function of the fraction frozen. The experimentally determined 50 % temperature depression is calculated by using a least-squares fit of the freezing-curve data over the range from the peak value to the 60 % frozen value. The 100 % frozen value is estimated to be that value when the freezing-curve temperature was depressed by approximately 10 m°C, or where the slope of the curve no longer becomes increasingly negative.

The calculated and the experimentally determined temperature depressions may be compared to confirm the overall purity of the metal sample in the fixed-point cell. Differences between the calculated and the experimentally determined temperature depressions may indicate an uncertainty in the use of Raoult's Law of dilute solutions (use of equation 1 assumes that during slow freezing all of the impurities remain in the liquid phase with no concentration gradient), an uncertainty in the extrapolation method chosen to derive the temperature depression, an uncertainty in the quantities of the impurities specified in the emission spectrographic assay, or that additional impurities were inadvertently added to the metal during the construction of the fixed-point cells. In estimating the expected temperature depression, Raoult's Law of dilute solutions is intended to provide a guideline and does not strictly apply.

3.2 Melting-point realization

After the metal sample was slowly and completely frozen in the above manner (≥10 h), the furnace temperature was set at about 1 °C above the freezing-point temperature to slowly melt the metal over an average time of 10 hours. Using an excitation current of 1 mA, thermometer readings were recorded continuously until the melting was complete. After each melting-point realization, the SPRT was measured at the TPW to calculate W(Sn). The experimentally determined 100% fraction melted value was calculated by using a least-squares fit of the melting-curve data over the range melted from the 20 % value to the 80 % value, assuming the fraction is directly proportional to the fraction of the duration of the melt.

For metal samples having a purity of <99.999 99 %, the use of a melting curve to determine the purity of the metal is complicated by the fact that the shape and range of a melting curve will depend on the history of the previous freezing of the metal in the fixed-point cell [7-9]. A slow freeze (≥10 h) causes the impurities to be segregated, which in turn causes a melting range indicative of the impurities. A fast freeze (<30 min) causes a homogenous mixture of the impurities within the metal sample, which in turn causes a small melting range. A fast freeze is realized by removing the fixed-point cell to ambient temperature and allowing the molten metal to freeze.

During a _slow_ melt of the metal sample, following a _slow_ freeze, two liquid-solid interfaces may be formed. One liquid-solid interface is next to the inner wall of the graphite crucible and the second liquid-solid interface is _near_ the graphite re-entrant well. This second liquid-solid interface is formed where the lower-purity metal solidified at the end of the previous _slow_ freeze. The lower-purity metal has a slightly lower freezing and melting temperature which may cause this second liquid-solid interface to form during the melt. [8, 10].

3.3 Direct comparison measurements

The second part of the certification was a direct comparison of the fixed-point cells under test with the laboratory standard fixed-point cell to determine their freezing-point temperatures relative to that of the reference cell. The three Sn cells containing the SRM metals were directly compared with the NIST laboratory reference cell Sn 88A. This was performed by realizing "simultaneous" freezes for the two cells in two separate but nearly identical furnaces and making three sets of alternate measurements, at equal time intervals from the start of the freezes, on their freezing-curve plateaus, using an SPRT. This ensures that the comparison measurements on the two cells were made at approximately the same liquid-solid ratio of the metal samples. Ideally, the equivalent temperature difference between measurements of each of the pairs would be identical. However, due to small differences in sample purity, only the first of the three pairs of measurements on the cells was used for the comparison. The other two pairs of measurements on the cells provided information on the progress of the freezes. The fraction of metal frozen during the first pair of measurements in the set of the three measurement pairs of the freezing-point temperature of each cell did not exceed 20 %. The SPRT was measured with excitation currents of 1 mA and 1.414 mA to permit extrapolation to zero-power dissipation (0 mA). Since the cells containing the SRM tin and the reference cell are of the same design and maintained at the same pressure, no corrections for pressure and hydrostatic head effects were necessary. Using an SPRT, each cell was measured three times during the direct comparison and this procedure was repeated two times. Following each set of direct comparison measurements, the SPRT was measured at the TPW to calculate W(Sn).

4. Analysis of Results

4.1 Analysis of SRM tin freezing-point curves

Figures 2 to 4 show the freezing curves for each of the three Sn fixed-point cells (the region of supercooling and recalescence are not shown, as the curves begin after the reinsertion of the thermometer). Using an excitation current of 1 mA, thermometer readings were recorded continuously until the freezing was complete. The average length of a freeze for the three Sn cells was 21.2 hours. The time-temperature relationships shown in the figures were calculated from the change in resistance of the SPRT during a freeze. For comparison purposes, the three freezing curves for each cell shown in the figures were normalized so that the maximum SPRT resistance obtained during the freezing-point realization is equivalent to the 0 m°C point on the graph. The calculated temperature depression when 50 % of the metal was frozen and the shapes of the freezing-curve plateaus were used to estimate and confirm the overall purity of the sample in each cell.

In most cases, impurities present in a high-purity Sn sample will cause a depression in the freezing-point temperature. Using the first cryoscopic constant of tin (0.003 29 K^{-1}) and the total mole fraction impurity (8.29 x 10^{-7} mol) calculated from the assay of the metal sample, equation 1 gives a calculated estimation of the depression in temperature of 0.25 m°C. As determined from the freezing curve plateaus in figures 2 to 4, the average estimated temperature depression from the extrapolated 0 % frozen metal (time of recalescence) to the 50 % frozen metal was 0.16 m°C, 0.13 m°C, and 0.13 m°C for Sn 97-1, Sn 97-2 and Sn 97-3, respectively. Thus, for the freezing-point cells containing the SRM tin, the experimentally determined temperature depressions are smaller than the calculated value. Differences between the calculated and the experimentally derived depressions may indicate either an uncertainty in using Raoult's Law of dilute solutions, an uncertainty in the extrapolation method chosen to determine the temperature depression, or an uncertainty in the impurities specified in the emission spectrographic assay.

Comparisons of the average temperature depressions when 50 % of the metal of the three fixed-point cells was frozen were made with that of Sn 88A. The average temperature depression from the extrapolated 0% frozen metal (time of recalescence) to the 50 % frozen metal was 0.07 m°C for Sn 88A. Using the first cryoscopic constant of tin and the total mole fraction impurity level (9.76×10^{-8} mol) calculated from the assay of the metal sample used in Sn 88A, equation 1 gives an estimation of the depression in temperature of 0.03 m°C. Since, the experimentally determined temperature depression for Sn 88A is slightly larger than the calculated value, this may indicate either an uncertainty in using Raoult's Law of dilute solutions, an uncertainty in the extrapolation method chosen to derive the temperature depression, or an uncertainty in the impurities specified in the emission spectrographic assay. The depressions in the freezing-point temperatures for the three cells relative to that of the reference cell, however, are consistent with purity differences of the SRM tin from that of the Sn contained in the reference cell.

4.2 Analysis of SRM tin melting-point curves

Figures 4 to 7 show the melting curves for each of the three Sn fixed-point cells. Using an excitation current of 1 mA, thermometer readings were recorded continuously until the melting was complete. The average length of a melt was 10.8 hours. The time-temperature relationships shown in the figures were calculated from the change in resistance of the SPRT during a melt. For comparison purposes, the three melting curves for each cell shown the figures were normalized so that the SPRT resistance corresponding to 50 % melted sample passes through the 0 m°C point on the graph. From the graphs, the average temperature ranges of the melting curves were 1.1 m°C, 1.3 m°C, and 1.1 m°C for Sn 97-1, Sn 97-2 and Sn 97-3, respectively.

While using a melting curve to determine the purity of the metal is difficult, an analysis of the difference in the liquidus-point temperatures obtained from a slow freeze (≥10 hours) and from a melt after a fast freeze may be performed. However, as an additional indicator of high-purity (≥ 99.9999 % pure) metal, the difference in the liquidus-point temperatures determined from the freezing (0 % frozen) and melting (100 % melted) curves of a fixed-point cell should be within 0.2 m°C [8]. The difference in the liquidus-point temperatures determined from the freezing and melting curves for the three Sn cells was about 0.2 m°C.

4.3 Analysis of direct comparisons

The second part of the certification was a direct comparison of the fixed-point cells under test with the laboratory standard fixed-point cell (Sn 88A) to determine their freezing-point temperatures relative to that of the reference cell. Figure 8 shows the results of the direct comparison of the three

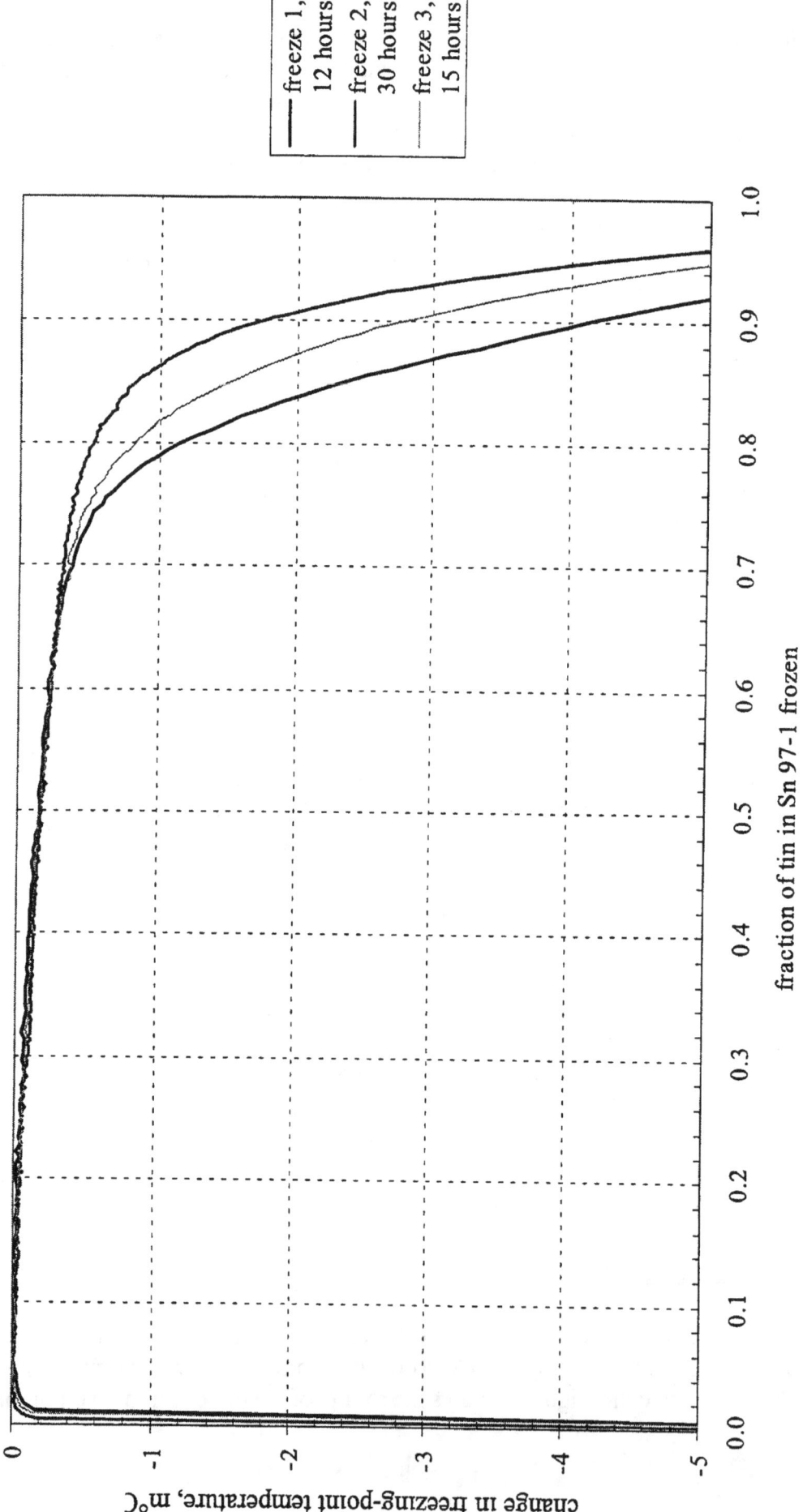

Figure 2. Three freezing curves for the tin fixed-point cell Sn 97-1 using the "induced inner freeze" preparation technique.

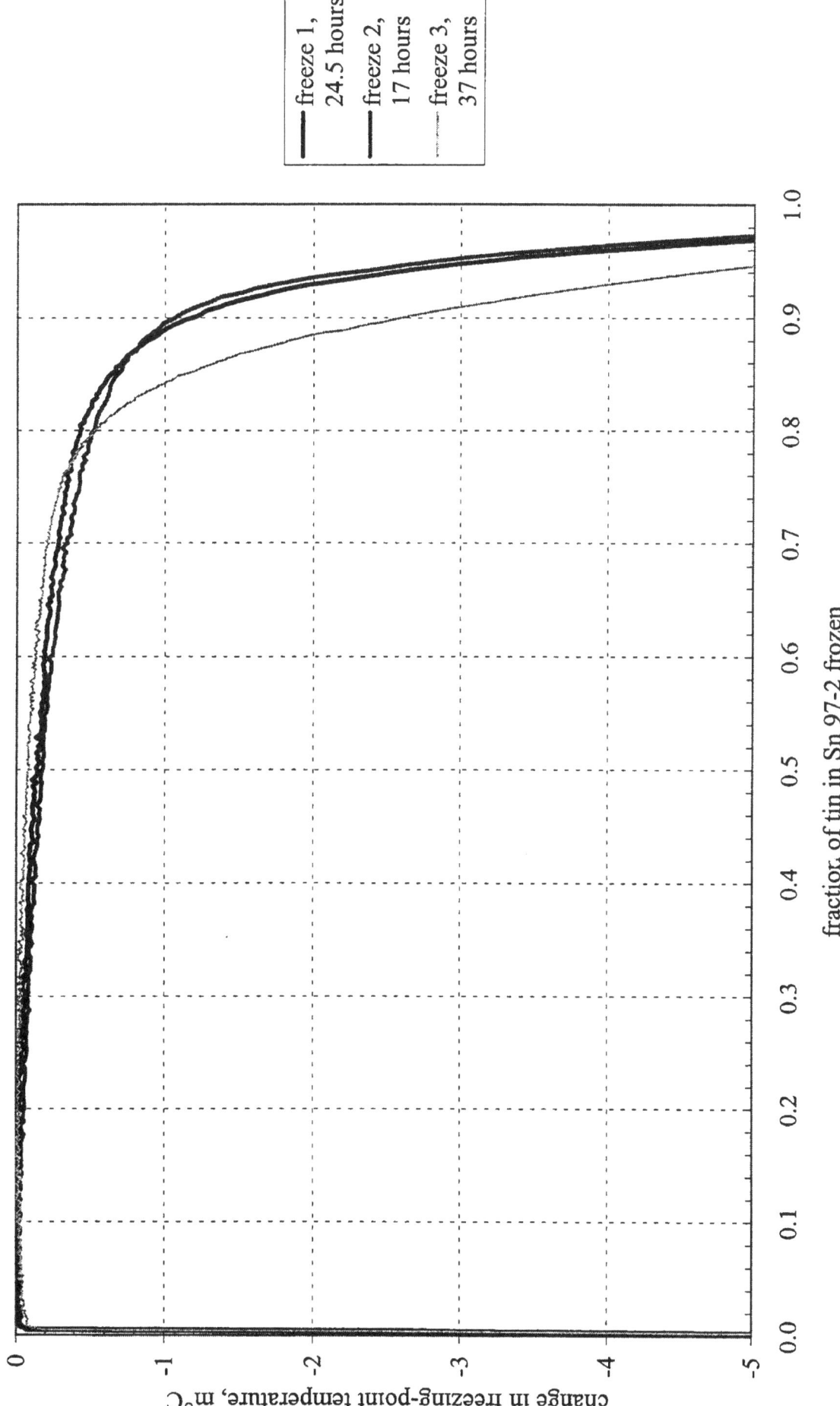

Figure 3. Three freezing curves for the tin fixed-point cell In 97-2 using the "induced inner freeze" preparation technique.

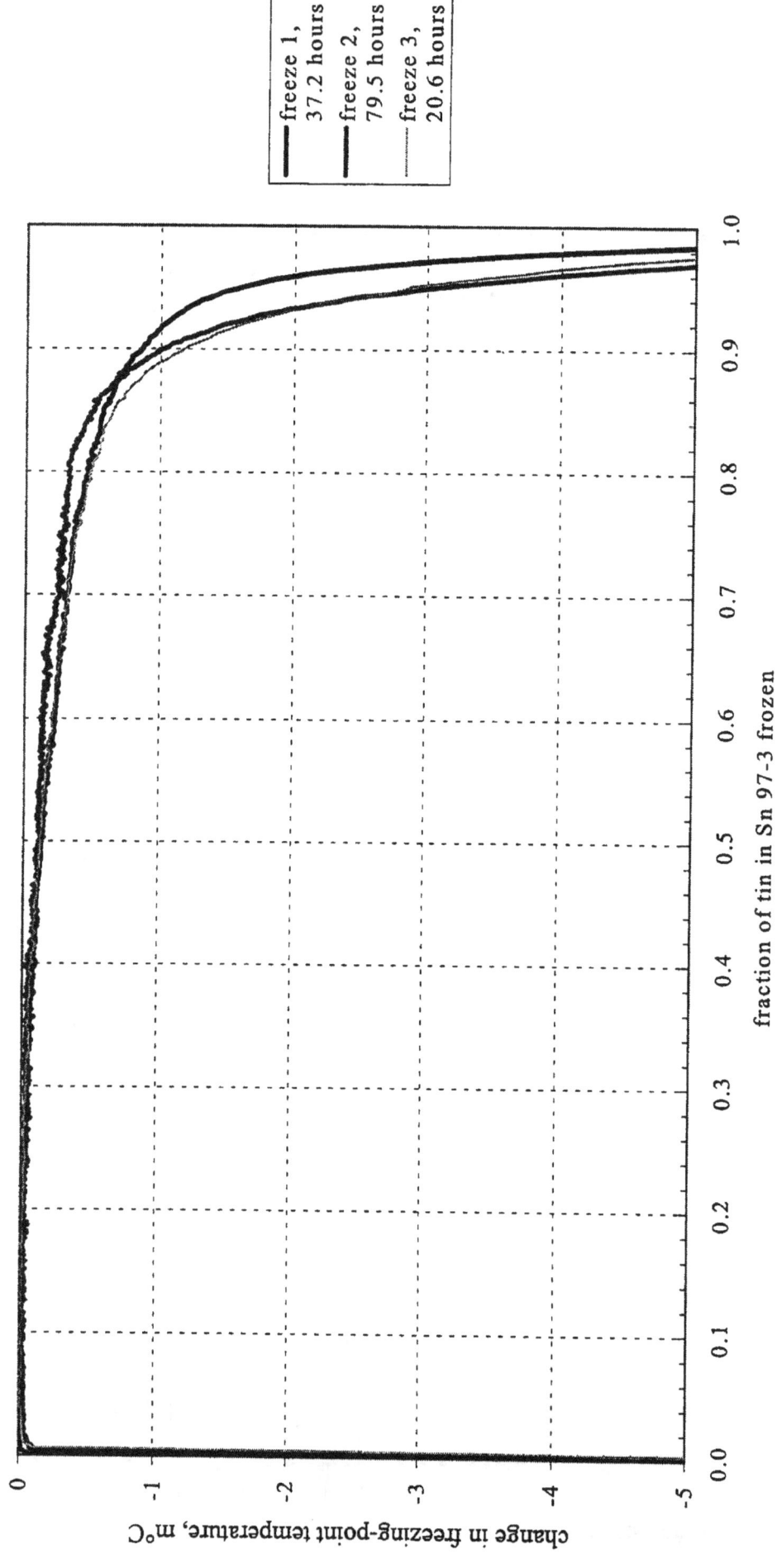

Figure 4. Three freezing curves for the tin fixed-point cell Sn 97-3 using the "induced inner freeze" preparation technique.

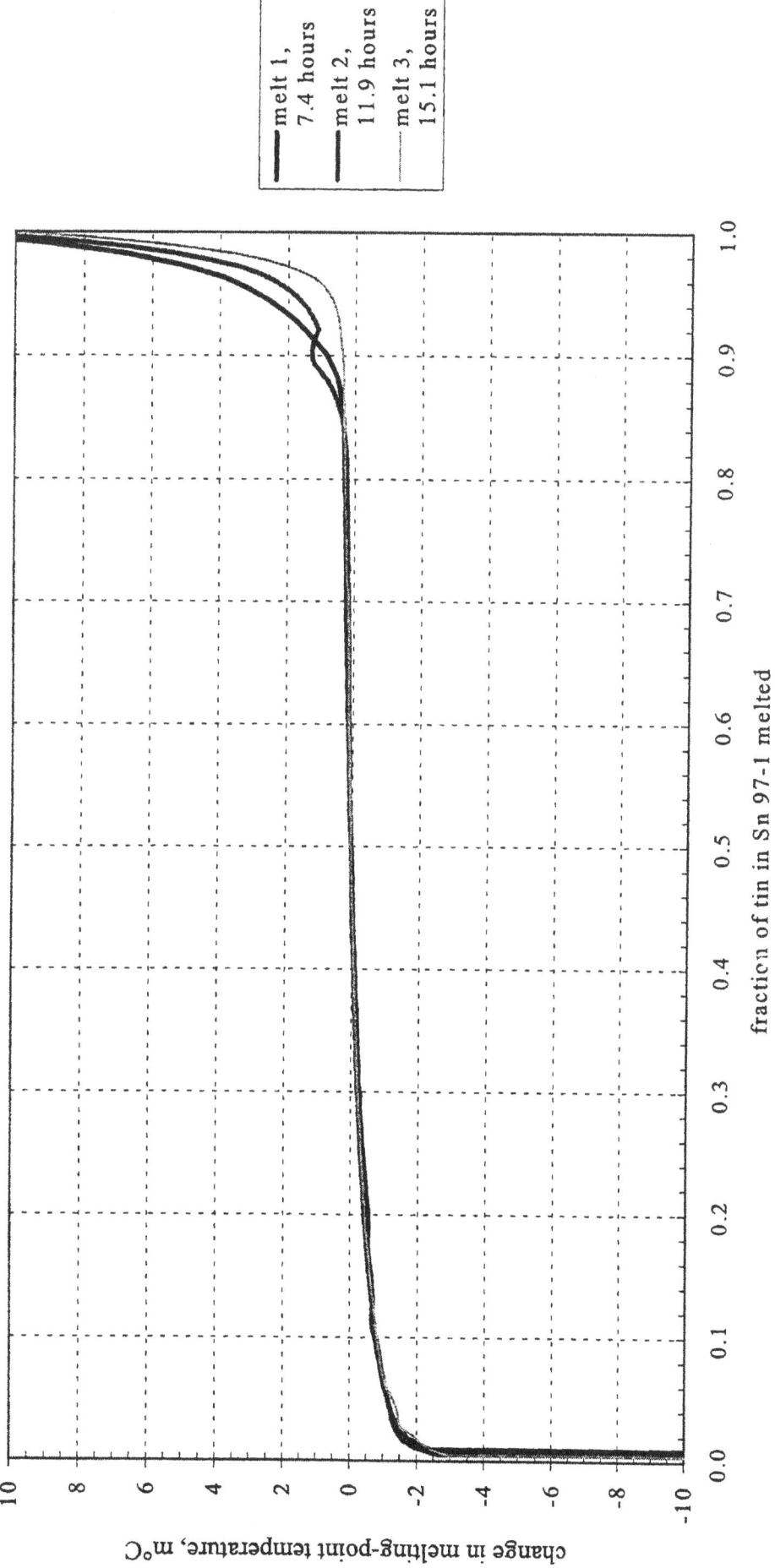

Figure 5. Three melting curves for the tin fixed-point cell Sn 97-1 following a slow freeze. Each melt followed the respective slow freeze of Figure 2.

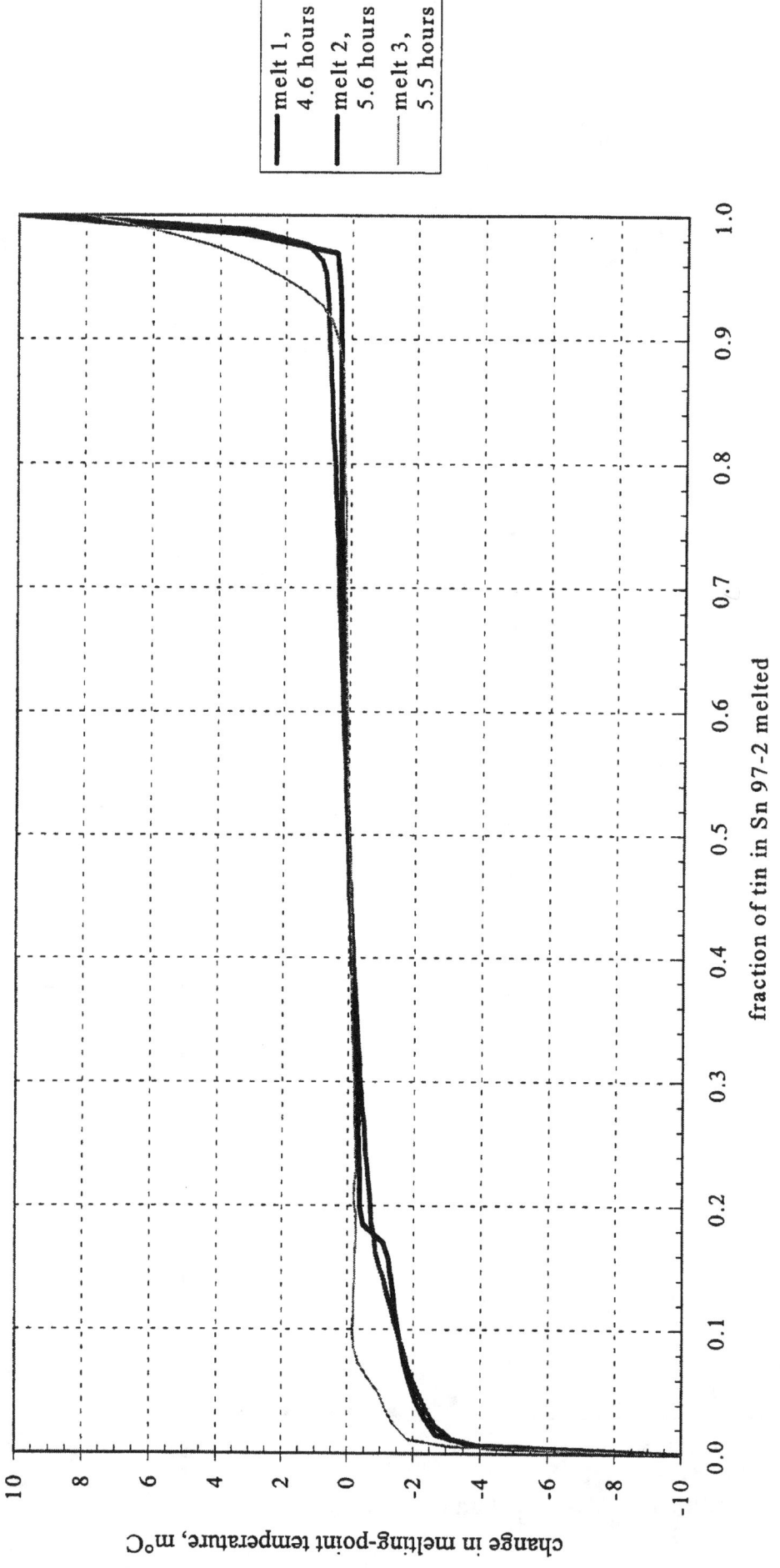

Figure 6. Three melting curves for the tin fixed-point cell Sn 97-2 following a slow freeze. Each melt followed the respective slow freeze of Figure 3.

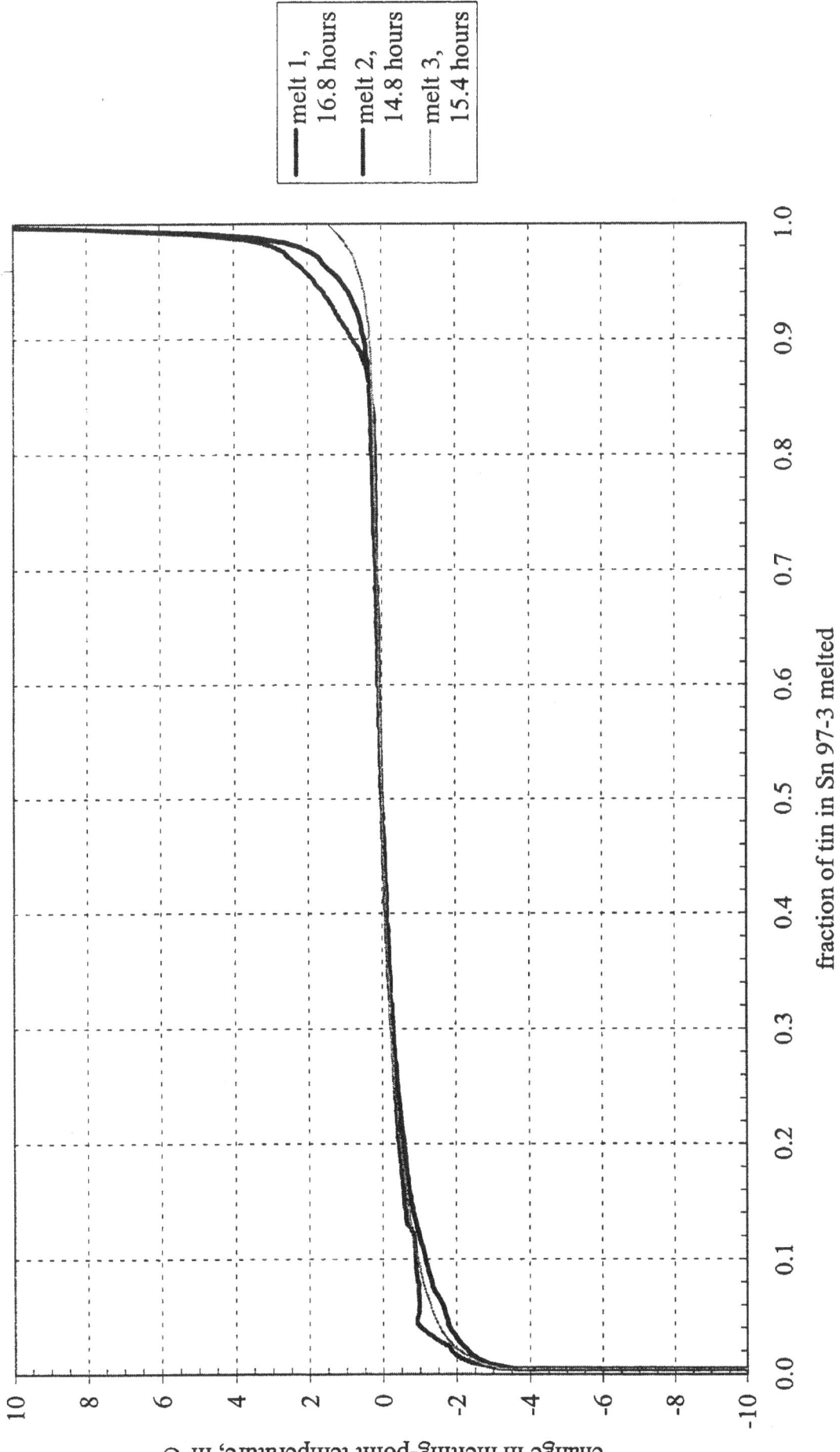

Figure 7. Three melting curves for the tin fixed-point cell Sn 97-3 following a slow freeze. Each melt followed the respective slow freeze of Figure 4.

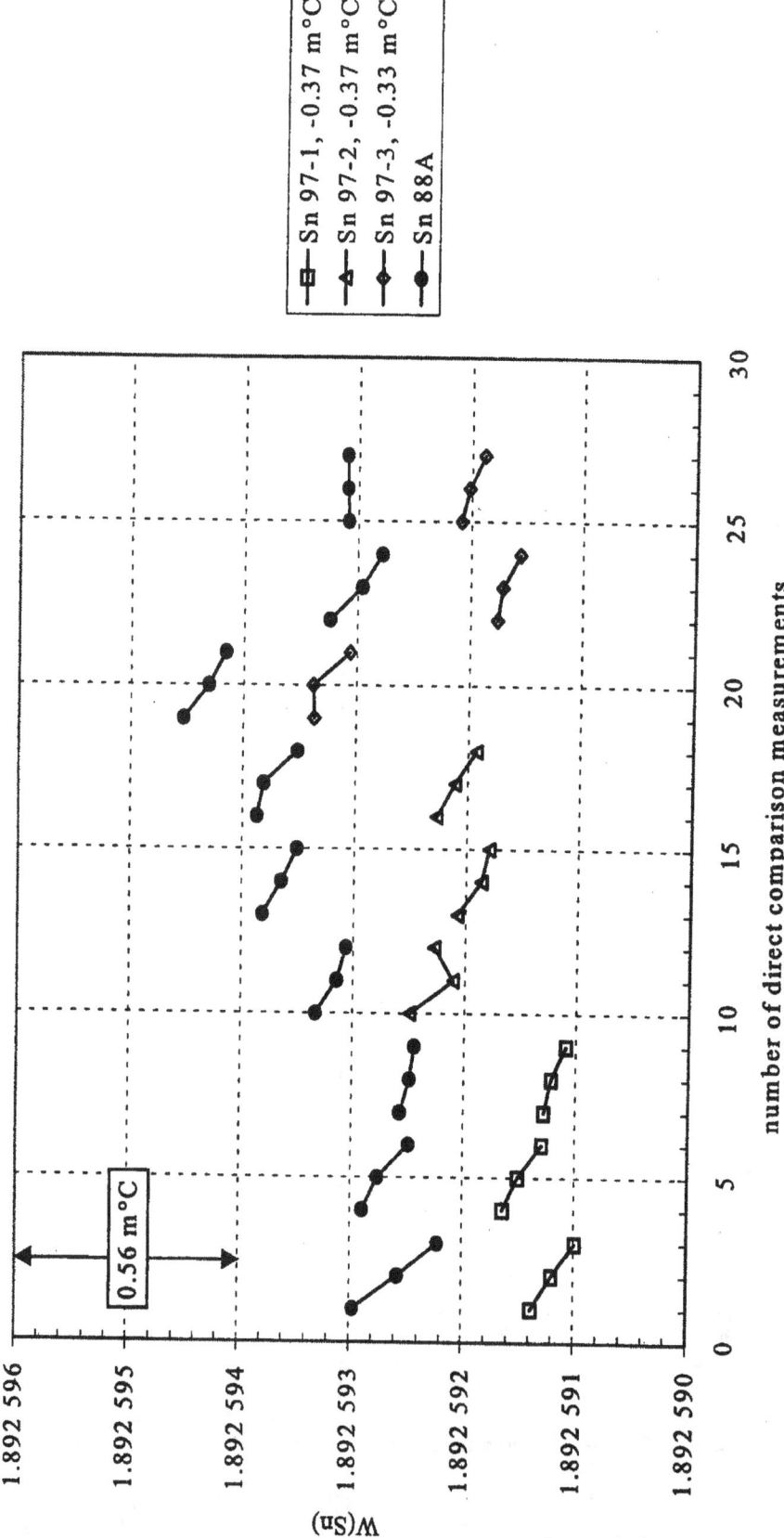

Figure 8. Direct freezing plateau comparison results of Sn 97-1, Sn 97-2 and Sn 97-3 with Sn 88A (laboratory standard). The differences shown in the legend represent the average temperature difference from the first readings of each direct comparison. The set of matching symbols (open and closed) are for the direct comparison measurements of Sn 97-X (where X is 1, 2 or 3) compared with Sn 88A. Each set of symbols (open or closed) connected by lines are for the direct comparison measurements made during simultaneous freezes.

Sn cells with the laboratory standard. The set of matching symbols are for the direct comparison measurements of Sn 97-X (where X is 1, 2, or 3) compared with Sn 88A. The average temperature difference of the first readings of each direct comparison showed that Sn 97-1, Sn 97-2 and Sn 97-3 were 0.37 m°C, 0.37 m°C and 0.33 m°C colder than the laboratory standard.

Each set of symbols connected by lines are for the direct comparison measurements made during the simultaneous freezes. These differences are consistent with the differences in impurity concentrations of the SRM metal and metal contained in the reference cell, as indicated by the emission spectrographic analysis.

5. Uncertainties

The expanded uncertainty U assigned to the measurements was calculated from the equation:

$$U = k\sqrt{s^2 + \Sigma u(i)^2} \qquad (3)$$

where k is the coverage factor, s is the Type A standard uncertainty and $u(i)$ is the estimated Type B standard uncertainty for each known component in the measurement process that cannot be directly measured [11, 12].

5.1 Uncertainty of direct comparison measurements

Many of the systematic effects in the measurement process cancel because the measurements being analyzed are from the direct comparisons of fixed-point cells. There were two possible known contributions to the Type A standard uncertainty; one from the instrumental measurements themselves and the second from the handling of the SPRT during transfer from cell to cell.

The calculated value of the Type A standard uncertainty for the direct comparison measurements was at most 0.07_4 m°C (n=6). The two contributions attributed to the Type A standard uncertainty were calculated to be at most 7.5 µ°C from the instrumentation and 74.2 µ°C from handling of the SPRT.

There were three known contributions to the Type B standard uncertainty in the direct comparison measurements. These were the uncertainty in the exact immersion depth of the SPRT due to the uncertainty in the position of the thermometer sensor during measurements, the uncertainty in the immersion depth of the thermometer due to the uncertainty in the exact fraction of the metal sample frozen, and the uncertainty in the adequacy of immersion of the thermometer to eliminate the thermometer stem conduction during the comparisons.

The Type B standard uncertainty from the three known contributions in the direct comparison measurements for <u>both</u> cells (SRM cell and the reference cell) was 0.00_9 m°C. The first contribution, coming from the uncertainty in knowing the exact immersion depth of the SPRT due to the uncertainty in the position of the thermometer sensor during measurements, gives an uncertainty of 1.3 µ°C. The second contribution, coming from the uncertainty in knowing the immersion depth of the thermometer due to the uncertainty in the exact fraction of the metal sample frozen, gives an uncertainty of 0.2 µ°C. The third contribution, coming from the uncertainty in the adequacy of immersion of the thermometer to eliminate the thermometer stem conduction during the comparisons, gives an uncertainty of 6.5 µ°C.

The expanded uncertainty ($k=2$) in the intercomparison measurements of the Sn fixed-point cells containing the SRM metal is 0.14_9 m°C.

5.2 Uncertainty assigned to SRM 741a

The expanded ($k=2$) uncertainty [11, 12] assigned to the SRM Sn freezing-point metal is 0.53 m°C. The Type A standard uncertainty of 0.15 m°C is the standard deviation of $W(t_{90})$ values of repeated measurements (n=100) of the laboratory standard Sn cell with a check SPRT [13]. The Type B standard uncertainty of 0.22 m°C is obtained from the average temperature difference between the three SRM cells and the laboratory standard Sn cell as determined from the direct comparison measurements of 0.37 m°C (see Section 4.3), and the uncertainty in those direct comparison measurements of 0.07 m°C (see Section 5.1). The two Type B standard uncertainties are considered to be a rectangular distribution and must be divided by $\sqrt{3}$.

6. Application

In assigning a temperature value to realizations of the tin freezing point, corrections must be applied for the depth of immersion (ℓ) of the thermometer sensing element below the surface of the metal ($dt/d\ell = 2.2 \times 10^{-3}$ °C/m) [1]. Also, if the pressure (p) over the cell during the measurements is not controlled at 101 325 Pa (1 standard atmosphere), a correction ($dt/dp = 3.3 \times 10^{-8}$ °C/Pa) must be made for the difference in pressure [1] from 101 325 Pa.

For those constructing their own tin freezing-point cell containing SRM 741a, it is necessary to confirm that the purity of the metal was maintained during construction of that cell. Using the methods described above, this confirmation is made by comparing the freezing and melting curves of the new cell with those shown in figures 2-10. As a continuing check on the overall purity of the tin metal contained in the fixed-point cell, melting and freezing curves should be obtained every six months and compared with those obtained previously. If a second tin fixed-point cell and a second furnace is available, then direct comparisons should be performed to verify the results from the freezing and melting curves.

7. Conclusions

The evaluation of SRM 741a, for use as an SRM tin freezing-point standard and as a replacement for SRM 741, has shown that the material is of high-purity (≥99.999 9%) and is acceptable for use as a defining fixed point of the ITS-90 and has been so certified. Using the methods of fixed-point cell fabrication and evaluation described above, the results from the evaluation of the three fixed-point cells containing the SRM metal show that the average temperature depression over the first 50% of a freezing-curve realization is not expected to exceed 0.16 m°C and the average temperature range of melting of the bulk material, following a slow freeze, is not expected to exceed 1.3 m°C. Plateau temperatures of the freezing curves for this material are expected to differ by less than 0.2 m°C from each other. SRM 741a has a higher purity (99.999 97%) and a smaller expanded uncertainty ($k=2$) of 0.53 m°C than that of the replaced SRM 741 (99.9999% pure and 1 m°C expanded uncertainty). A copy of the certificate for SRM 741a is given in appendix B.

8. References

[1] H. Preston-Thomas, "The International Temperature Scale of 1990 (ITS-90)," *Metrologia*, Vol. 27, pp. 3-10, (1990); *ibid*. p. 107.

[2] G.T. Furukawa, J.L Riddle, W.R. Bigge and E.R. Pfeiffer, "Application of Some Metal SRM's as Thermometric Fixed Points," NBS SP260-77, 140 pp. (1982).

[3] G.F. Strouse, "NIST Implementation and Realization of the ITS-90 Over the Range 83 K to 1235 K. Reproducibility, Stability, and Uncertainties," *Temperature. Its Measurement and Control in Science and Industry*, Edited by J.F. Schooley, Vol. 6, pp. 169-174, (American Institute of Physics, New York, 1992).

[4] E.H. McLaren, "The Freezing Points of High Purity Metals as Precision Temperature Standards," Canadian J. of Phys., **35**, pp. 1086-1106 (1957).

[5] B.W. Mangum and G.T. Furukawa, "Guidelines for Realizing the International Temperature Scale of 1990 (ITS-90)," NIST Technical Note 1265, 190 pp. (1990).

[6] S. Glasstone, *Thermodynamics for Chemists*, p. 322, (D. Van Nostrand Co., Inc., New York, 1947).

[7] B.W. Mangum, P. Bloembergen, M.V. Chattle, P. Marcarino and A.P. Pokhodun, "Recommended Techniques for Improved Realization and Intercomparisons of Defining Fixed Points", Report to the CCT by WG1, Comité Consultatif de Thermométrie, 19e Session (1996).

[8] Working Group 1 of the Comite Consultatif de Thermométrie (Mangum, B.W., Bloembergen, P., Chattle, M.V., Fellmuth, B, Marcarino, P., and Pokhodun, A.I., "On the International Temperature Scale of 1990 (ITS-90) Part II: Recommended Techniques for Optimal Realization of the Defining Fixed Points of the Scale that are Used for Contact Thermometry," submitted for publication.

[9] Working Group 1 of the Comite Consultatif de Thermométrie (Mangum, B.W., Bloembergen, P., Chattle, M.V., Fellmuth, B, Marcarino, P., and Pokhodun, A.I., "On the International Temperature Scale of 1990 (ITS-90) Part I: Some Definitions, Metrologia, 1997, Vol. 34, pp. 427-429.

[10] G.T. Furukawa, private communication.

[11] ISO, *Guide to the Expression of Uncertainty in Measurement*, International Organization for Standardization, Geneva, Switzerland (1993).

[12] B.N. Taylor and C.E. Kuyatt, "Guidelines for Evaluating and Expressing the Uncertainty of NIST Measurement Results," NIST Technical Note 1297, 17 pp. (1993).

[13] G.F. Strouse and W.L. Tew, "Assessment of Uncertainties of Calibration of Resistance Thermometers at the National Institute of Standards and Technology," NISTIR 5319, 16 pp. (1994).

9. Appendix A

Johnson Matthey emission spectrographic assay of tin metal for SRM 741a

JOHNSON MATTHEY ELECTRONICS
East 15128 Euclid Avenue, Spokane, Washington 99216
Tel.(509) 922-8781 / Fax.(509) 922-8652

CERTIFICATE OF COMPLIANCE AND ANALYSIS

CUSTOMER : National Institute of Standards & Technology
P.O. NO. : 40NANB708807

DATE : December 12, 1996
CONTRACT : 110967-002
LOT : M6755
QUANTITY : 24 Kg

DESCRIPTION : Shot

CUSTOMER SPECIFICATION	JOHNSON MATTHEY RESULTS
69 Grade Tin	Ag .1 Ca .1 Si .1

COMMENTS: Trace analysis by emission spectrograph.
Results in ppm wt.

This is to certify that this product has been made in accordance with Customer P.O. Number, Part No., and Material Specification listed above.

_____ Dec. 12, 1996
QA REPRESENTATIVE   DATE

10. Appendix B

Certificate for SRM 741a: Tin Freezing-Point Standard.

National Institute of Standards & Technology

Certificate of Analysis

Standard Reference Material® 741a

Tin Freezing-Point Standard

Certified Freezing-Point Temperature: (231.928 ± 0.000 53)°C

International Temperature Scale of 1990 (ITS-90)

This Standard Reference Material (SRM) is intended primarily for use as one of the defining fixed points of the International Temperature Scale of 1990 (ITS-90) [1]. The certified value of 231.928 °C ± 0.000 53 °C is the temperature assigned to the fixed point of SRM 741a. The fixed point is realized as the plateau temperature (or liquidus point) of the freezing curve of the slowly frozen high purity tin. SRM 741a consists of 200 g of tin in the form of millimeter size teardrop shot sealed in an argon atmosphere in a plastic bottle.

Based on samples tested, the temperature range of melting of this current lot of material is not expected to exceed 0.0013 °C. Temperatures of freezing curve plateaus (see Figure 2) for samples of this material are expected to differ by not more than ± 0.0002 °C from each other and by not more than ± 0.000 53 °C from the ITS-90 assigned temperature.

An expanded uncertainty ($k=2$) of 0.000 53 °C is assigned to the freezing-point temperature of SRM 741a. The Type A standard uncertainty component of 0.000 15 °C is the standard deviation of $W(t_{90})$ values of repeated measurements of the laboratory standard tin cell with a check Standard Platinum Resistance Thermometer (SPRT) [2]. The Type B standard uncertainty of 0.000 22 °C is obtained from the temperature difference between the three SRM cells and the laboratory standard tin cell as determined from the direct comparison measurements of 0.000 37 °C, and the uncertainty in those direct comparison measurements of 0.000 07 °C. The two Type B standard uncertainties are considered to be a rectangular distribution and must be divided by $\sqrt{3}$.

The tin for this SRM is of high purity, with the total of all elements that affect the freezing-point temperature being 0.3 mg/kg, resulting from 0.1 mg/kg of silver, 0.1 mg/kg of calcium, and 0.1 mg/kg of silicon.

Expiration of Certification: The certification of this SRM is valid indefinitely within the measurement uncertainties specified, provided the SRM is used in accordance with the instructions given in the Notice and Warnings to Users section of this certificate. The certification is nullified if the SRM is damaged, contaminated, or modified.

Source of Material: The tin metal (Lot M6755) for this SRM was obtained from Johnson Matthey Company of Spokane, Washington.

Temperature measurements of the fixed-point cells were performed by G.F. Strouse of the NIST Process Measurements Division and N.P. Moiseeva of D.I. Mendeleyev Research Institute of Metrology, St. Petersburg, Russia.

The support aspects involved in the preparation, certification, and issuance of this SRM were coordinated through the Standard Reference Materials Program by J.C. Colbert.

Gaithersburg, MD 20899
Certificate Issue Date: 2 September 1998

Thomas E. Gills, Chief
Standard Reference Materials Program

Notice and Warnings to Users: Because any handling of high purity material is apt to introduce contamination, this SRM is provided in "shot" form in order to minimize the need for handling during freezing-point cell construction. Nevertheless, every possible effort should be made to maintain the purity of this SRM through the use of polyethylene gloves while handling. Also, a clean laboratory environment is essential.

Instructions for Use: In assigning a temperature value to realizations of the tin freezing point for calibration purposes, corrections must be applied for the average depth of immersion (ℓ) of the thermometer sensing element below the surface of the metal ($dt/d\ell = 2.2 \times 10^{-3}$ °C/m). Also, if the pressure (p) over the cell during the measurements is not controlled at $1.013\,25 \times 10^2$ kPa (1 standard atmosphere), a correction, $dt/dp = 3.3 \times 10^{-8}$ °C/Pa, must be made for the difference in pressure.

Certification Testing: The thermal tests for the certification of this SRM were performed on three fixed-point cells prepared in a manner similar to that described in reference [3]. Each cell contains approximately 1071 g of tin obtained from randomly selected bottles of lot M6755.

The freezing points were prepared using the recommended "induced inner freeze" method. Due to the deep supercool of as much as 25 °C below the freezing-point temperature of high purity tin, the technique for the realization of the tin freezing point is different from that of the other ITS-90 freezing-point metals. The freezing point was achieved by heating the cell overnight to approximately 5 °C above the freezing-point temperature and then setting the furnace to a temperature of about 0.5 °C below the freezing-point temperature of the metal and monitoring the temperature of the metal with the check thermometer. When the tin had cooled to its freezing-point temperature, the freezing-point cell with the check thermometer was removed from the furnace until the start of recalescence. At the beginning of that recalescence, the fixed-point cell was placed back into the furnace, the check thermometer was removed and two fused-silica glass rods were inserted three minutes each into the reentrant well of the cell to induce an inner solid-liquid interface. Finally, the "cold" thermometer was reinserted into the cell and, after equilibrium was obtained, the measurements were started. After equilibrium was established, the temperature of the plateau on the freezing curve was found to vary no more than $\pm\,0.000\,16$ °C during the first 50 % of the duration of the freeze. Three freezing curves obtained under such conditions are shown in Figure 1 (the region of supercooling and recalescence is not shown, as the curves begin after the reinsertion of the thermometer); a sample of the data is plotted at greater resolution in Figure 2.

After the metal was slowly and completely frozen in the above manner, the furnace was set to a temperature of about 1 °C above the freezing-point temperature to slowly melt the metal over an average time of about 10 h. Thermometer readings were recorded continuously until the melting was complete. Three melting curves obtained under such conditions are shown in Figure 3; some of the same data are plotted at greater resolution in Figure 4.

Following the freezing and melting curve measurements, the plateau temperature of a freezing curve of the test cell was compared directly with that of the standard tin freezing-point cell maintained by the NIST Platinum Resistance Thermometer Calibration Laboratory, using a 25.5 Ω SPRT. The method of direct comparison is described in detail in reference [5].

During the freezing and melting curve measurements, an inert environment of argon gas at 101 325 Pa $\pm$ 27 Pa was maintained in the cells.

The electronic measurement equipment included an ASL F18[1] resistance ratio bridge, operating at a frequency of 30 Hz, and a temperature controlled Tinsley 5685A 100 Ω reference resistor. This reference resistor was maintained at a temperature of (25.000 ± 0.010) °C. Freezing curve and melting curve measurements were made with an excitation current of 1 mA. Direct comparison measurements of the thermometer resistance were conducted at two excitation currents, 1 mA and $\sqrt{2}$ mA, with a 25.5 Ω SPRT, to allow analysis of the results at zero power dissipation. A computer controlled data acquisition system was used to acquire the ASL F18 bridge readings through the use of an IEEE-488 bus.

---

[1] Certain commercial materials and equipment are identified in order to adequately specify the experimental procedure. Such identification does not imply a recommendation or endorsement by the National Institute of Standards and Technology, nor does it imply that the materials or equipment are necessarily the best available for this purpose.

REFERENCES

[1] Preston-Thomas, H., "The International Temperature Scale of 1990 (ITS-90)," Metrologia **27**, pp. 3-10 (1990); Metrologia **27**, p. 107, (1990).

[2] Strouse, G.F. and Tew, W.L., "Assessment of Uncertainties of Calibration of Resistance Thermometers at the National Institute of Standards and Technology," NISTIR 5319, 16 pages, (1994).

[3] Furukawa, G.T., Riddle, J.L., Bigge, W., and Pfeiffer, E.R., "Standard Reference Materials: Application of Some Metal SRM's as Thermometric Fixed Points," Natl. Bur. Stand. (U.S.), Spec. Publ. 260-77, 140 pages, (1982).

[4] Mangum, B.W. and Furukawa, G.T., "Guidelines for Realizing the International Temperature of 1990 (ITS-90)," NIST Tech. Note 1265, 190 pages, (1990).

[5] Mangum, B.W., Pfeiffer, E.R., and Strouse, G.F. (NIST); Valencia-Rodriguez, J. (CENAM); Lin, J.H. and Yeh, T.I. (CMS/ITRI); Marcarino, P. and Dematteis, R. (IMGC); Liu, Y. and Zhao, Q. (NIM); Ince, A.T. and Cakiroglu, F. (UME); Nubbemeyer, H.G. and Jung, H.J. (PTB), "Intercomparisons of Some NIST Fixed-Point Cells with Similar Cells of Some Other Standards Laboratories," Metrologia **33**, (1996).

Users of this SRM should ensure that the certificate in their possession is current. This can be accomplished by contacting the SRM Program at: Telephone (301) 975-6776 (select "Certificates"), Fax (301) 926-4751, e-mail srminfo@nist.gov, or via the Internet http://ts.nist.gov/srm.

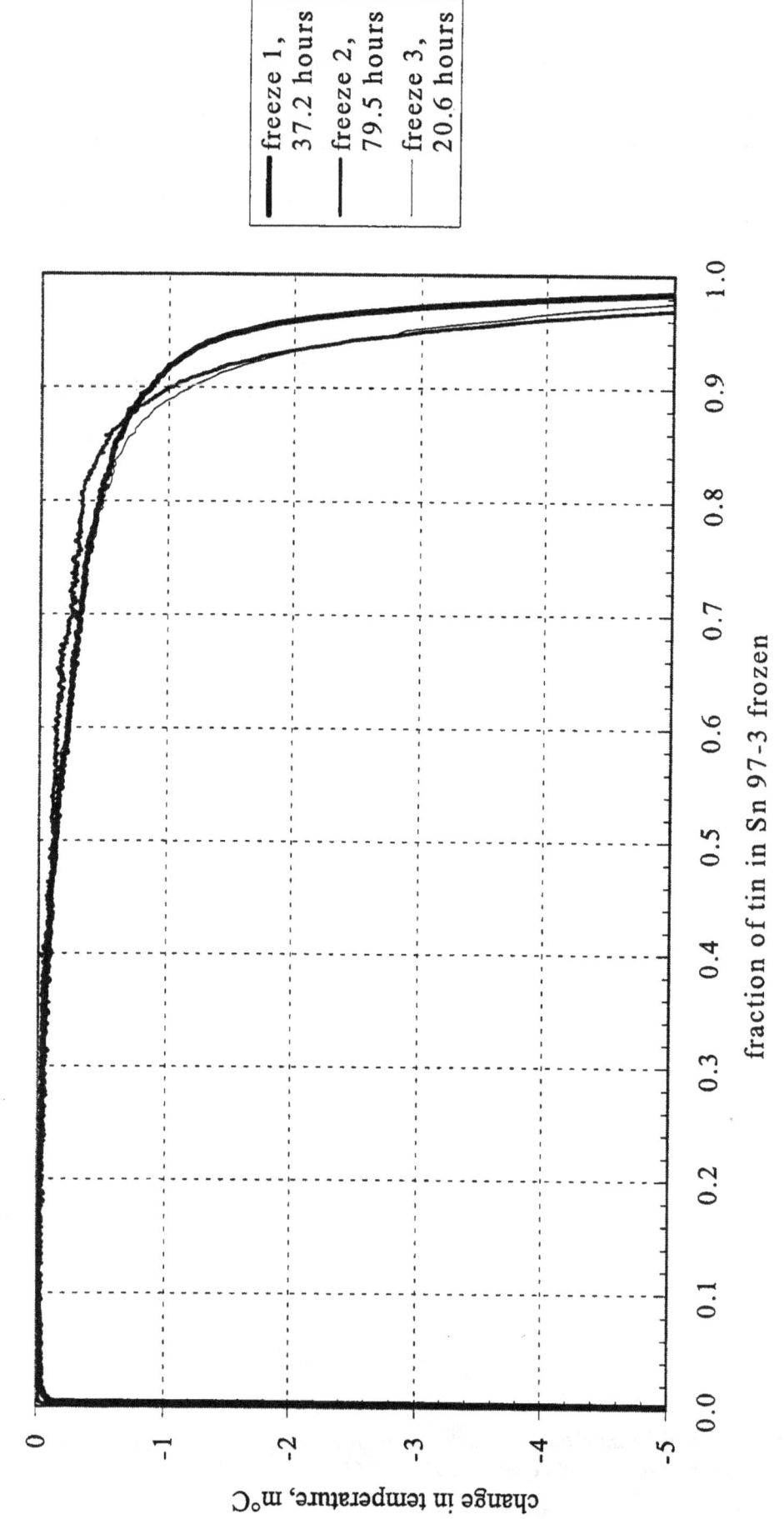

Figure 1. Three freezing curves for SRM 741a using the "induced inner freeze" preparation technique.

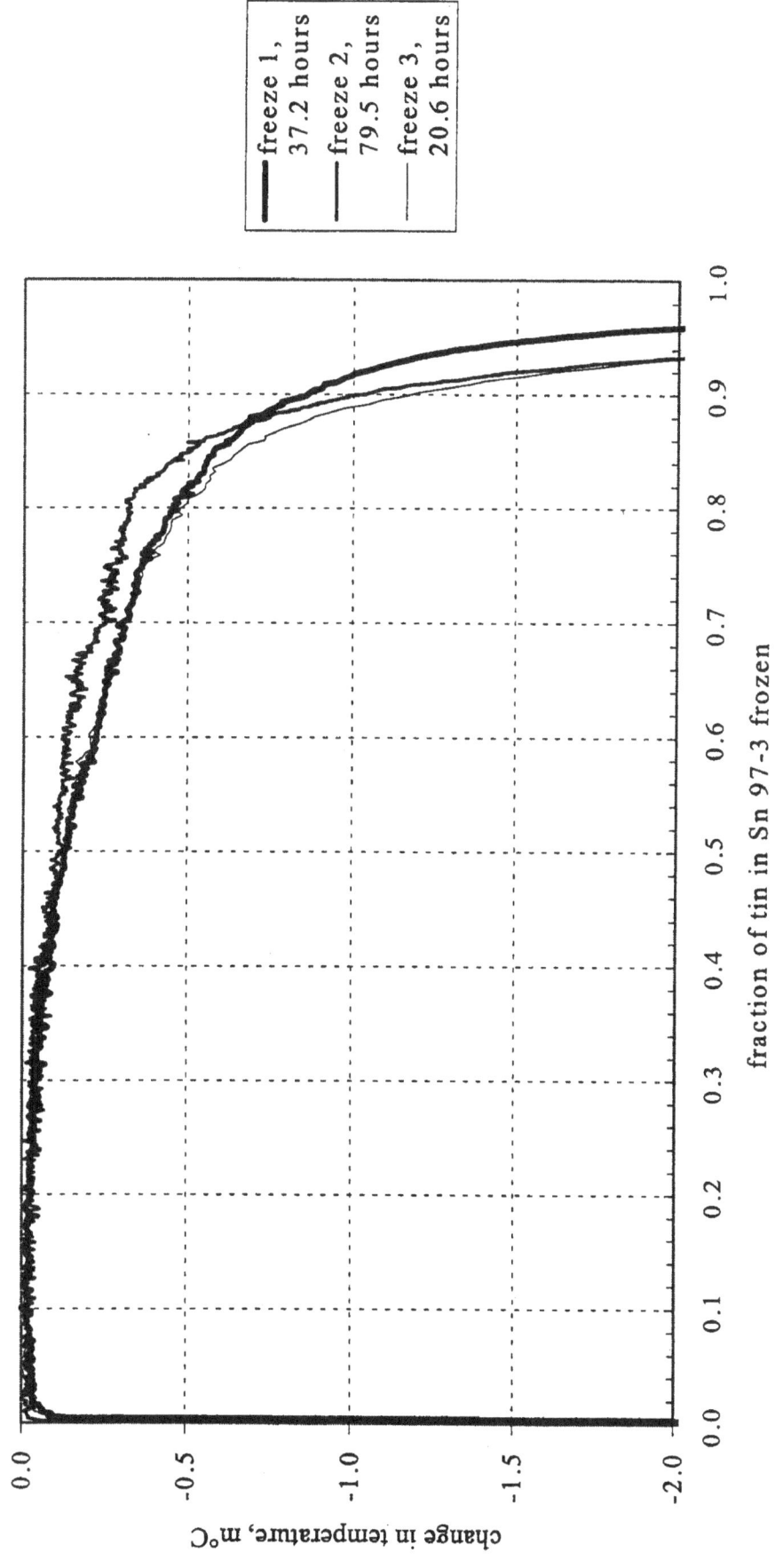

Figure 2. The freezing plateau regions of Figure 1 at greater resolution.

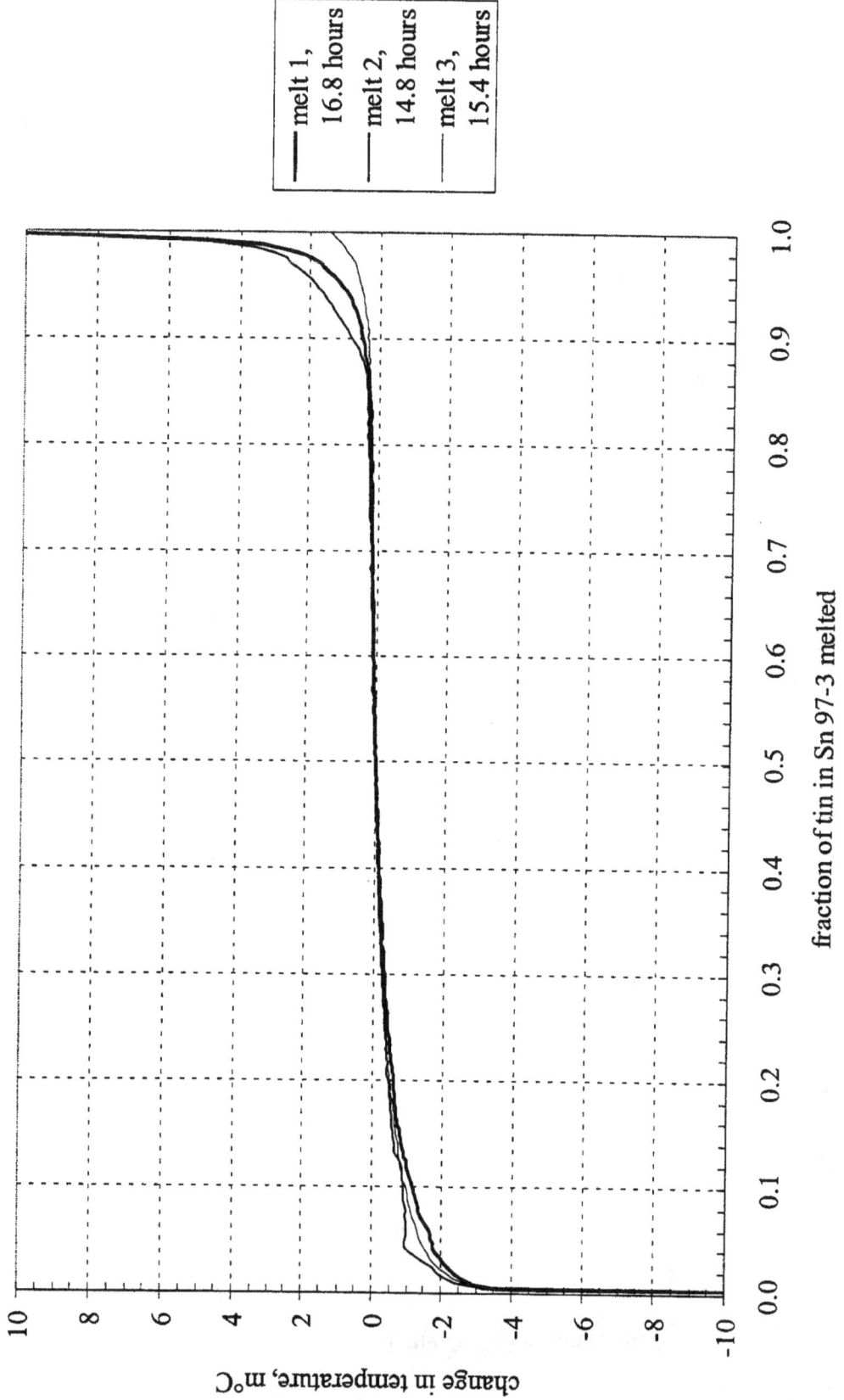

Figure 3. Three melting curves of SRM 741a tin following a slow freeze. Each melt followed the respective slow freeze of Figure 1.

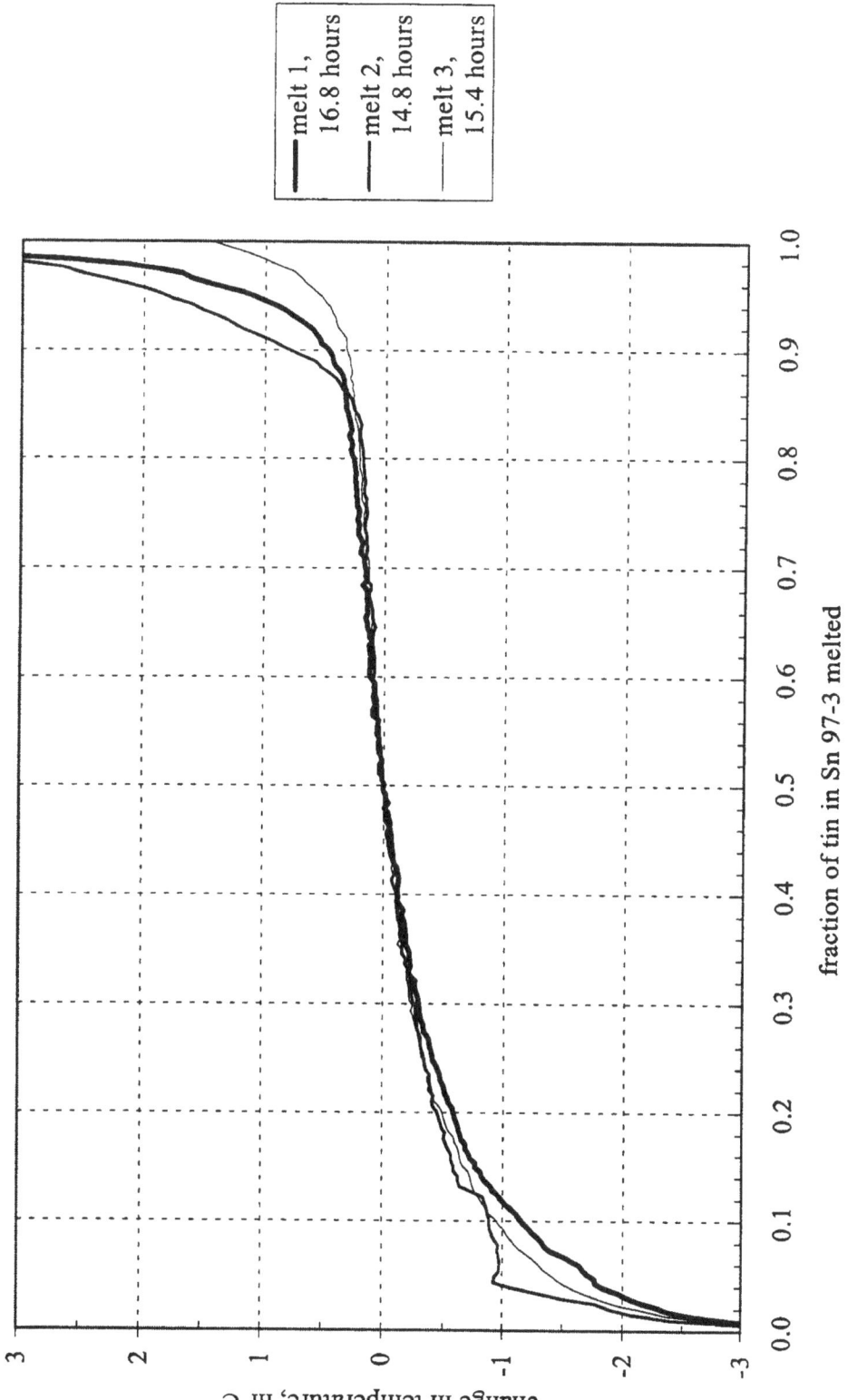

Figure 4. The melting plateau regions of Figure 3 at greater resolution.

www.ingramcontent.com/pod-product-compliance
Lightning Source LLC
Chambersburg PA
CBHW081807170526
45167CB00008B/3366